YOUR AMAZING BRAIN
AND HOW IT WORKS

For Emily, whose way of looking at the world helps me to understand my own. – S.E.

For Otis & Lenny. Be kind and always believe in yourselves. I'm SO proud of you both. Love Dad xx. – T.B.F.H.

Text and illustrations copyright © b small publishing 2025

First Sky Pony Press Edition 2025

All rights reserved. No part of this book may be reproduced in any manner without the express written consent of the publisher, except in the case of brief excepts in critical reviews or articles. All inquiries should be addressed to Sky Pony Press, 307 West 36th Street, 11th Floor, New York, NY 10018.

Sky Pony Press books may be purchased in bulk at special discounts for sales promotions, corporate gifts, fund-raising or education purposes. Special editions can also be created to specifications. For details, contact the Special Sales Department at Skyhorse Publishing, 307 West 36th Street, 11th Floor, New York, NY 10018 or info@skyhorsepublishing.com.

Sky Pony Press® is a registered trademark of Skyhorse Publishing, Inc.®, a Delaware corporation.

Visit our website at www.skyhorsepublishing.com.

Publisher: Sam Hutchinson

Creative director: Vicky Barker

Editorial: Alice Harman

10 9 8 7 6 5 4 3 2 1

LCCN Control Number: 2025005229

Manufactured in China on FSC-certified paper, June 2025

This product conforms to CPSIA 2008

ISBN: 978-1-5107-8383-6
Ebook ISBN: 978-1-5107-8466-6

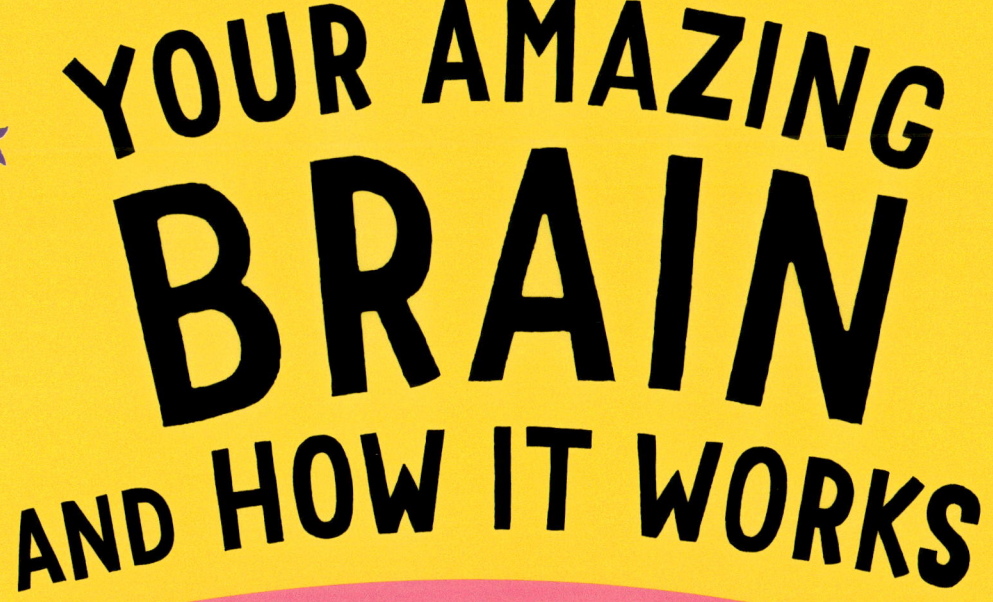

YOUR AMAZING BRAIN AND HOW IT WORKS

An Inclusive Guide for Kids

written by
SCOTT EVANS
THE READER TEACHER

expert advice from **DR. RACHEL S.H. WILLIAMS**

illustrated by
THE BOY FITZ HAMMOND

designed by
VICKY BARKER

Sky Pony Press
New York, NY

CONTENTS

6-7   ABOUT THIS BOOK

SCIENCE

8-9   BRAINY BEINGS

10-11   YOUR BRAIN'S BEGINNING

12-13   THE BOSS OF YOUR BODY

14-15   UNDERSTANDING THE WORLD

16-17   TRAIN YOUR BRAIN

18-19   FACING BRAIN CHALLENGES

SOCIETY

20-21   NEURODIVERSITY

22-23   PAST, PRESENT, AND FUTURE

24-25   DISCRIMINATION

26-27   IS NEURODIVERGENCE A DISABILITY?

TAKING ACTION

28-29   MENTAL HEALTH MATTERS

30-31   MINDFUL MAINTENANCE

32-33   WHAT CAN YOU DO?

34-36   GLOSSARY AND USEFUL WORDS

ABOUT THIS BOOK

This book is a celebration of ALL brains! Read on to find out about...

...the fascinating journey your brain takes as it develops during your life.

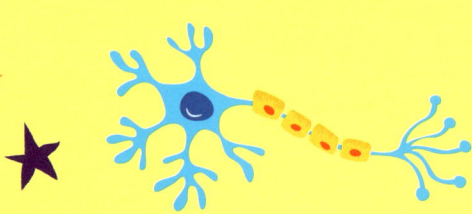

...the amazing things happening inside your brain, which make you who you are.

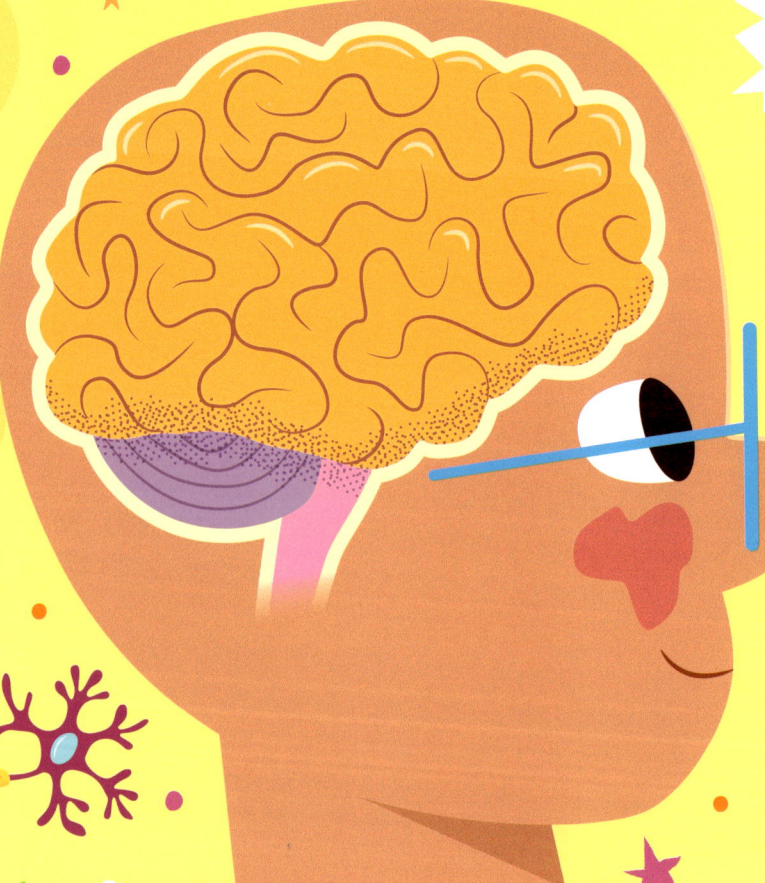

...how even though our brains may look similar, each and every one is unique.

BRAINY BEINGS

Your brain is *incredible*. It works non-stop to keep you breathing, moving, thinking, feeling, learning, and so much more. And the best part? It's unique to you, making it a V.I.B—a Very Individual Brain!

However, humans aren't the only ones with brains. Other animals have them too, in all shapes and sizes.

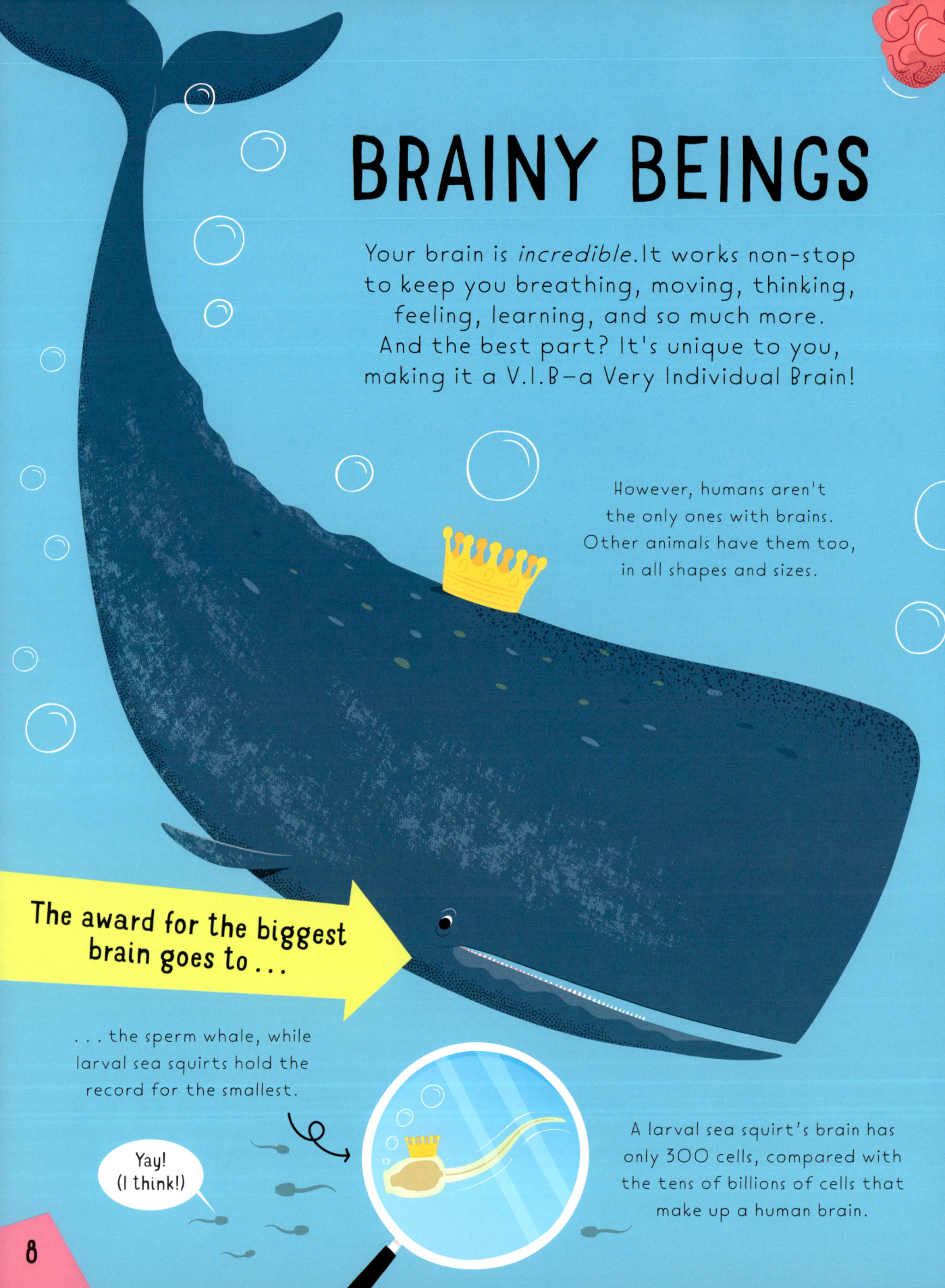

The award for the biggest brain goes to...

...the sperm whale, while larval sea squirts hold the record for the smallest.

Yay! (I think!)

A larval sea squirt's brain has only 300 cells, compared with the tens of billions of cells that make up a human brain.

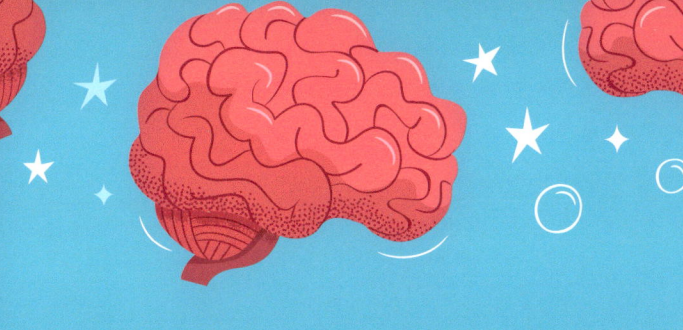

Brain-body mass ratio tells us how big an animal's brain is in comparison to its body.

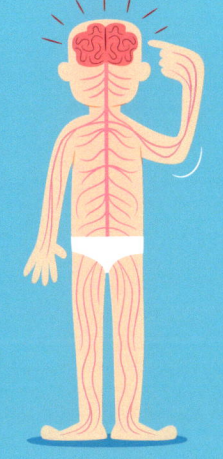

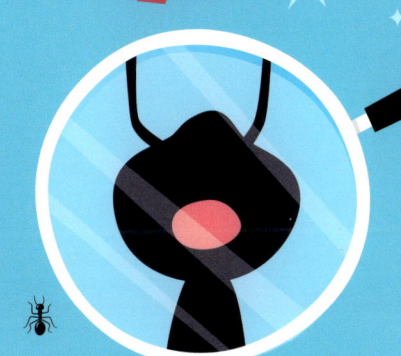

The sperm whale has a **1:5100** ratio, which means their brains are big but their bodies are massive!

Humans have a **1:40** brain-body mass ratio.

Small ants have a **1:7** ratio. Big brains for their tiny bodies!

But brace yourself to meet the bony-eared assfish (yes, that's its real name), whose brain is less than 0.1% of its weight.

Having a bigger brain doesn't always mean being smarter, though. And some animals, such as jellyfish and starfish, don't have a brain at all! Neither do trees and plants, but they have parts that behave like brains to help control how they grow, take in water, and respond to light.

But enough about them—let's focus on you! Curious about your brain before you were born? Get ready for some mind-boggling secrets . . .

YOUR BRAIN'S BEGINNING

Close your hands into fists and put them together. That's how big your brain is inside your head, right now!

But it hasn't always been this size. It started out super small. So small that you'd need a microscope to see it!

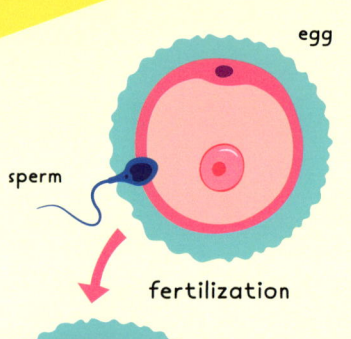

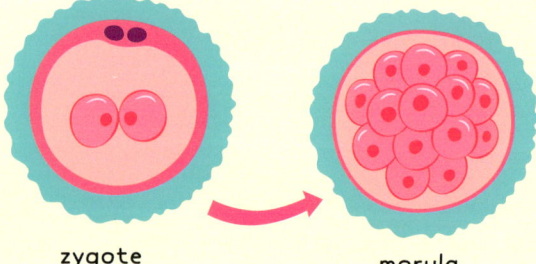

Next, the morula becomes a blastocyst, as it takes in fluid and forms a protective sac around the developing cells.

When an egg and a sperm come together, they make a zygote—a tiny cell that quickly divides into a clump of 16-32 cells called a morula. This is named after the Latin word for "mulberry." Can you see why?

mulberry

Once the blastocyst finds a safe spot in the uterus, it begins to form three layers—the ectoderm, mesoderm, and endoderm—that shape different parts of the human body.

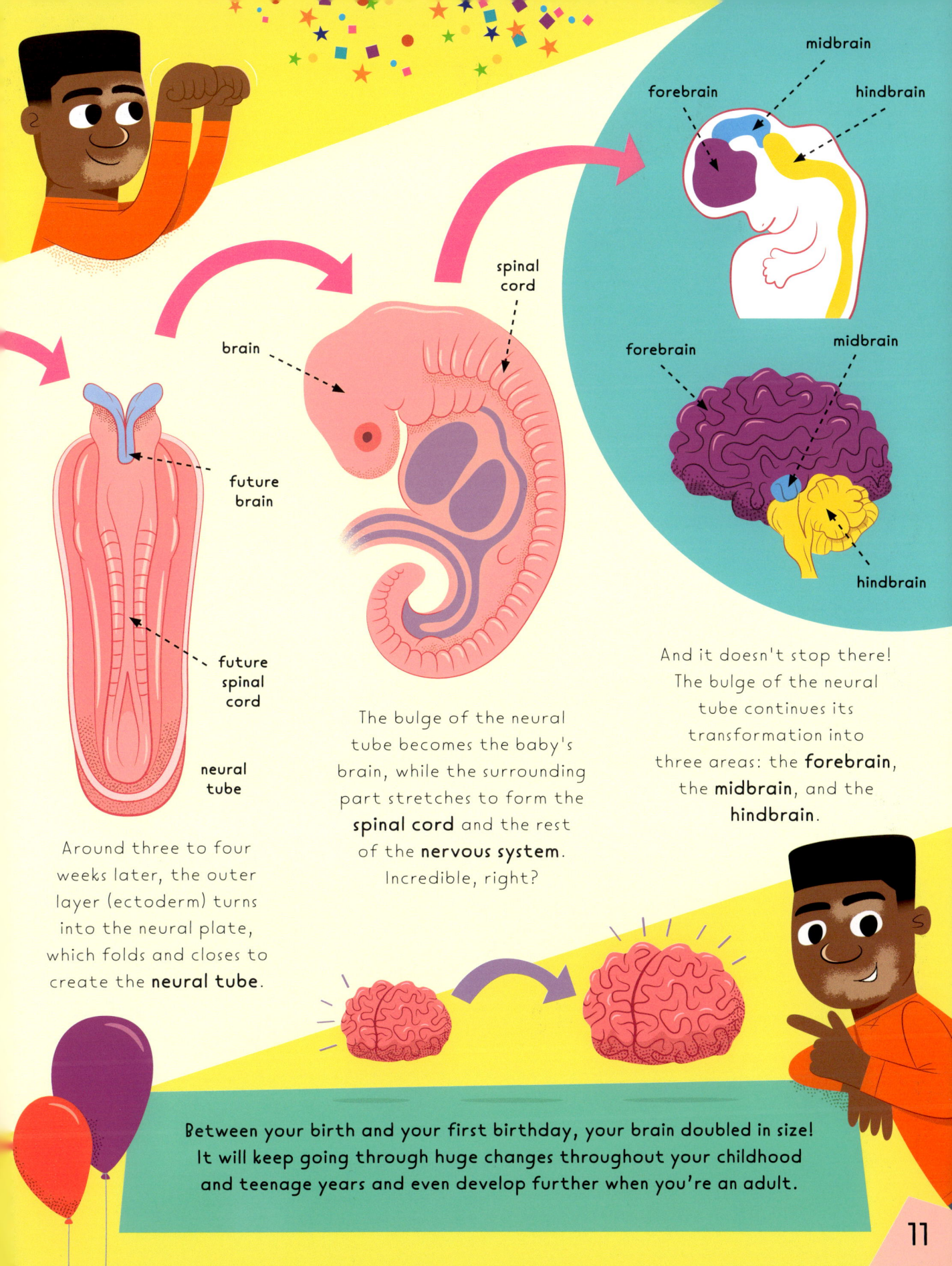

THE BOSS OF YOUR BODY

From thinking and blinking to talking and walking, your brain controls everything you do. The three main parts of your brain—the cerebrum, the cerebellum, and the brainstem—each have their own specific role.

The cerebrum is the largest part, and is split into sections called lobes. Each lobe has some main functions, although many overlap:

The **frontal lobe** is where much of your planning and decision-making takes place. It also forms your personality, controls your movements, and manages your memory, emotions, and (partly) speech.

In the middle, the **parietal lobe** processes information about touch, temperature, pain, your surroundings, writing, and (partly) language.

On the sides, the **temporal lobes** help with hearing, understanding spoken and sign language, recognizing objects, and storing memories.

At the back, the **occipital lobe** makes sense of what your eyes see, including movement and color.

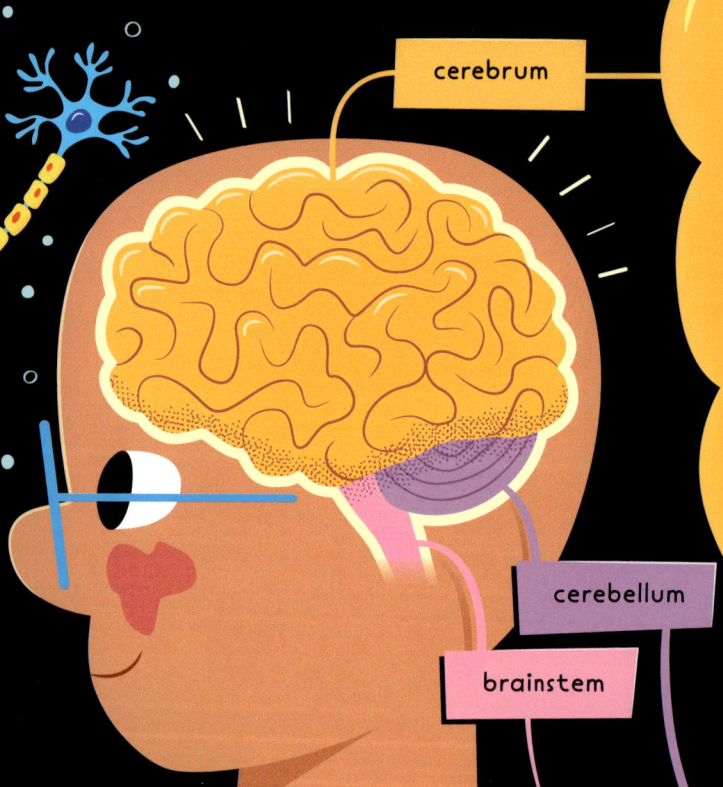

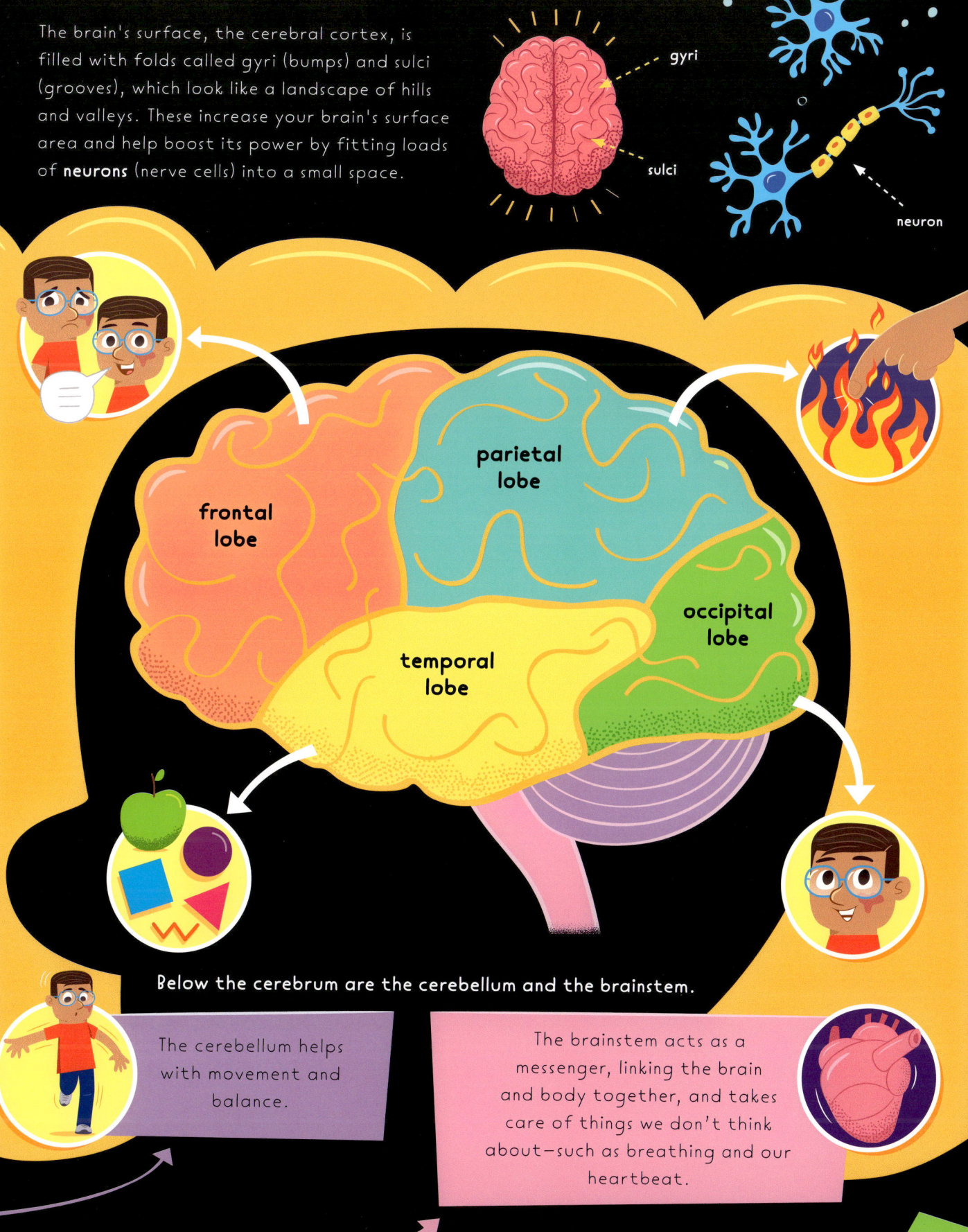

UNDERSTANDING THE WORLD

Your brain works a lot like a sponge, soaking up information from your surroundings and everything that you do. This includes what you read, see, hear, taste, and the people you meet, along with the things you learn in school and the toys and games you play with.

Different parts of your brain (see page 12) process all of this information to help you understand and respond to the world. This can play an important part in shaping how you think and act.

Your family are your first teachers because they share their beliefs and behaviors with you from a young age. Sometimes, you might act like them without realizing it!

Remember... you are your own person!

Your brain continues to grow and change throughout your life, and you have the power to develop your own beliefs as you get older. You may pick up new ideas from friends, but you don't need to follow what they do either—you can choose what's right for you.

The world is filled with other influences such as books, websites, TV shows, movies, advertisements, and social media. While these can be engaging, entertaining, and educational, they can also sometimes spread incorrect information and harmful ideas.

That's why it's important to be curious, ask questions, and check that information comes from reliable and accurate sources, like scientists and detectives do!

TRAIN YOUR BRAIN

Did you know that your brain can keep changing, even when you're an adult? This is called neuroplasticity. It means your brain can reshape itself by forming new connections and losing old ones throughout your life.

Inside your brain, over 86 billion **neurons** are at work. Neurons send and receive signals that carry messages about everything you think, feel, and do. These signals can travel along pathways to **synapses**—points of contact between neurons— at up to 268 miles per hour. That's faster than a Formula 1® race car!

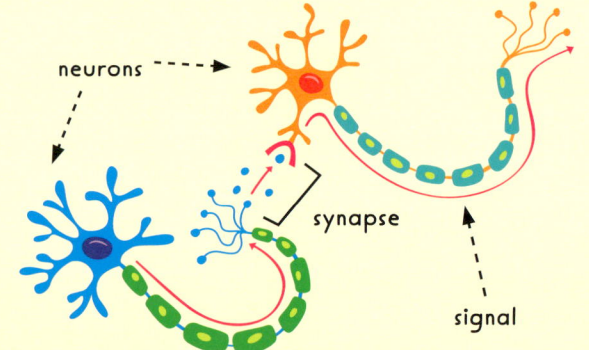

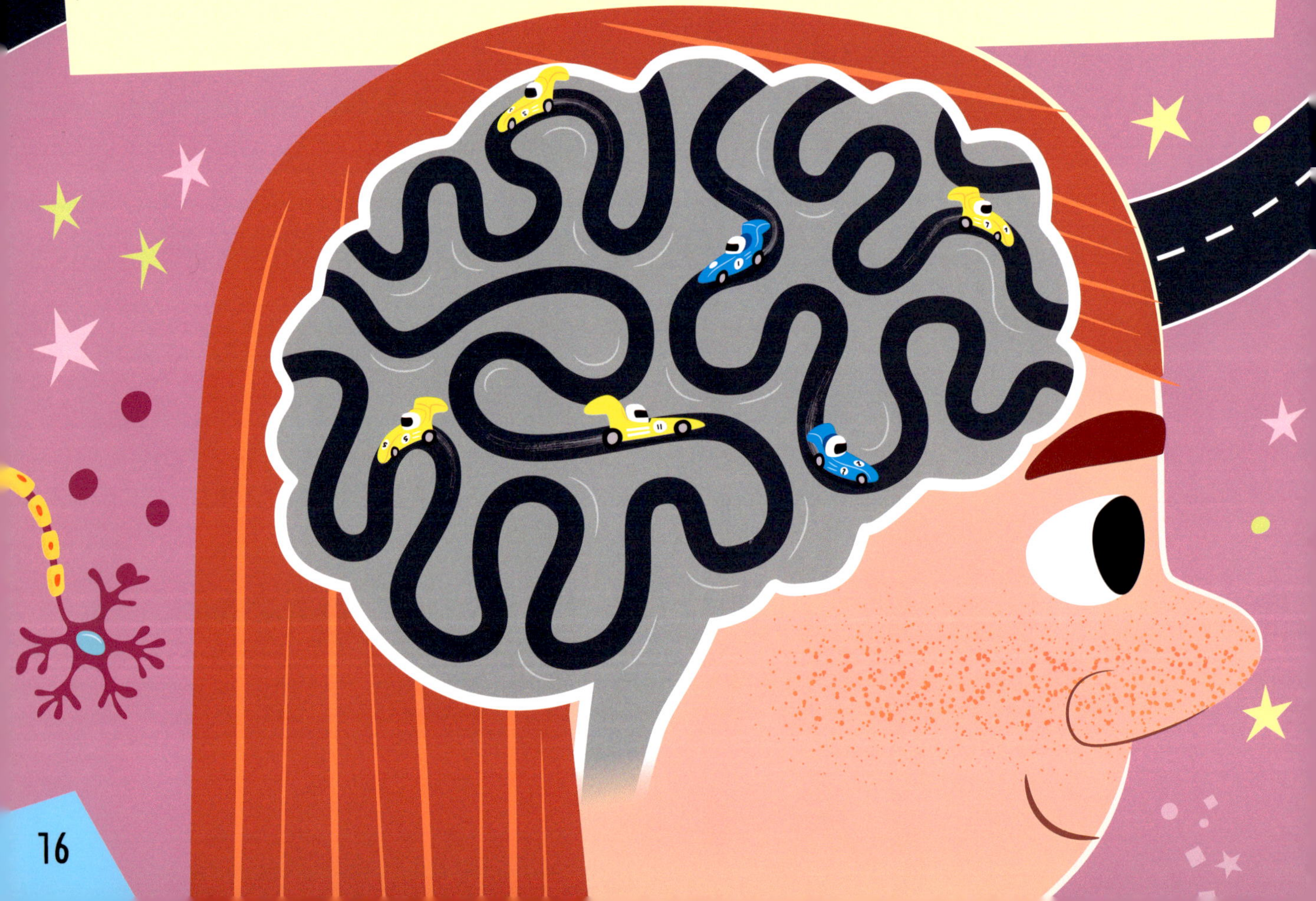

What you learn and experience in life can strengthen connections between your brain cells and shape and speed up the pathways in your brain. This can happen at synapses, as more electrical impulses travel between certain neurons. It can also happen through a process called myelination, which adds a coating to these pathways that makes them stronger and smoother.

Neuroplasticity means that the more you practice something, the better you'll become at it. This takes time and continues happening throughout your life.

Learning isn't just about getting things right—mistakes are like stepping stones to success. Whenever you make a mistake, your brain learns from it and finds another way to improve.

So, don't be afraid to make mistakes—it's your brain's way of saying, "Let's do even better!"

FACING BRAIN CHALLENGES

Different things can happen to our brains. They can sometimes experience changes that cause conditions called neurological disorders. Some of these conditions are present when we're born, while others can develop as we grow older.

Spina bifida occurs when the neural tube (see page 11) doesn't completely close around the developing spinal cord during early pregnancy.

Symptoms of **cerebral palsy** vary from person to person, affecting movement and muscle coordination. This condition results from brain damage that has occurred before, during, or after birth.

Epilepsy involves bursts of electrical activity in the brain, leading to seizures—short periods of losing control of certain body functions.

Motor neurone disease (MND) causes nerve cells that control muscles to break down. This causes muscle weakness and issues with movement, speech, and breathing.

Parkinson's disease affects the control of body movements and gets worse over time. It can lead to symptoms such as shaking, slow movements, and losing a sense of smell.

A **stroke** happens when blood flow to the brain is blocked, making parts of the brain and the rest of the body stop working properly. It's important to get help quickly if someone has a stroke.

A **concussion** is a brain injury that can happen when we get a hard knock to our head, for example in a fall or while playing contact sports such as football. Repeat concussions can be particularly dangerous.

Dementia makes remembering and understanding things difficult. This can make everyday tasks tougher. It can affect any adult, not just older people.

People with neurological disorders often use courage and determination to face challenges in their daily lives. We can all create positive change by listening to their needs and supporting them personally and in our communities.

NEURODIVERSITY

Imagine if we all thought, felt, and learned in exactly the same ways—the world would be a lot less interesting! Neurodiversity is the idea that our brains all work in very individual ways. This makes each of us unique and helps us all bring something special to the world we live in.

Everyone has their own way of thinking, learning, communicating, and processing information. **Neurotypical** people's brains tend to work in ways that society expects, while **neurodivergent** people's brains often work differently to this. There's no right or wrong way, and everyone deserves the same respect, understanding, and inclusion.

There are many forms of neurodivergence, including . . .

- autism
- dyspraxia
- Tourette's syndrome
- dyslexia
- dysgraphia
- synesthesia
- ADHD
- dyscalculia
- tics
- hyperlexia

. . . and more.

It's common for neurodivergent people to have multiple **diagnoses**, known as co-occurring conditions. These can be mental health conditions or other neurodivergent traits.

For example, an **autistic** person might also have **hyperlexia** and **OCD**.

In a world designed for neurotypical people, it can be tough for those who think differently. Some neurodivergent people might hide their true thoughts and feelings. This is called masking. While it can help people fit in, it can also make them feel tired, alone, and misunderstood.

Despite these challenges, neurodivergent people show incredible and unique strengths such as creativity, innovation, and **resilience**.

When we embrace and increase our understanding of neurodiversity and include everyone, our world becomes a friendlier and happier place for all of us.

PAST, PRESENT, AND FUTURE

For thousands of years, humans have studied the brain using all sorts of different methods. But our knowledge of neurodiversity and **neurodivergence** is something we've only started to learn about more recently.

Let's take a closer look at how things have changed.

In the past

Throughout history, not everyone understood how unique our brains are. If someone's brain didn't work as society expected, many people thought there was something medically wrong with them that needed "fixing." This made it difficult for neurodivergent people to get the understanding and support they needed and deserved.

Jim Sinclair is often described as the founder of the autism rights movement. Since the 1980s, Sinclair has spread the message that **autistic** people don't need to be "cured."

In the present

We now know more about neurodiversity than ever before. We're aware that every brain is one of a kind and that neurodivergent people are important and valuable just the way they are. Events such as Neurodiversity Celebration Week, World Autism Awareness Day, and **ADHD** Awareness Month recognize neurodiversity and neurodivergence.

However, there's still much more to be done! For example, many neurodivergent people—especially those from certain groups, such as Black women and girls—have a much harder time getting a **diagnosis** and proper support. This is partly because most research and resources have typically focused on white men and boys.

In the future

Advances in science and technology, such as brain scanning, are helping us to better understand neurodiversity and neurodivergence.

Teaching **neuroinclusivity** in schools and workplaces—that is, looking at how society and individuals can adapt to better include neurodivergent people—is also helping to shape a more welcoming world.

How could you help build a brighter, more neuroinclusive future?

DISCRIMINATION

Discrimination is when people are treated differently because of their age, **disability**, race, religion, gender, sex, or **sexual orientation**. These are known as protected characteristics, which means it's illegal to treat someone unfairly because of them.

Everyone's identity is complex, and people often have different combinations of protected characteristics.

Although **neurodivergence** isn't a protected characteristic yet, it's very important to make sure that neurodivergent people aren't treated unfairly.

Direct discrimination—for example, when someone isn't given a job because of their disability—is often easy to spot. But indirect discrimination can be less obvious at first, as it typically involves treating everyone "the same," which means not taking people's specific needs into account. This puts some people at an unfair disadvantage.

All discrimination makes life harder for neurodivergent people. It can limit the opportunities available to them and be harmful to their wellbeing. This is why it's so important to work together to stop discrimination and to speak out against **stereotypical** (wrong and overly simple) ideas that can lead to it.

I'm **autistic**, but that doesn't mean I can't experience emotions or make friends, as many might believe. I have big feelings and enjoy spending time with my friends!

Lots of people think I can't sit still or concentrate because of my **ADHD**. Yep, I've got tons of energy, but I can put it into what I love, which is when I achieve amazing things!

Being **dyslexic**, it's really frustrating when some people say I can't be interested in reading and writing. I actually love books!

IS NEURODIVERGENCE A DISABILITY?

Neurodivergence is sometimes described as "hidden" or "non-visible." People often use these words when talking about **disabilities**, which might lead them to think that neurodivergence is a disability.

Did you know around 15-20% of people on Earth are considered neurodivergent? That's about 1.5 billion people, nearly 1 in every 6!

Just because differences aren't always immediately visible, it doesn't mean that they're not there. They can still impact someone's life.

Some countries around the world view neurodivergence as a disability, but many others don't. Do you think neurodivergence should be considered a disability everywhere? Why or why not?

In the UK, some neurodivergent people like to wear a sunflower lanyard to let others know they need additional support in public places. How would you help if you saw someone wearing one?

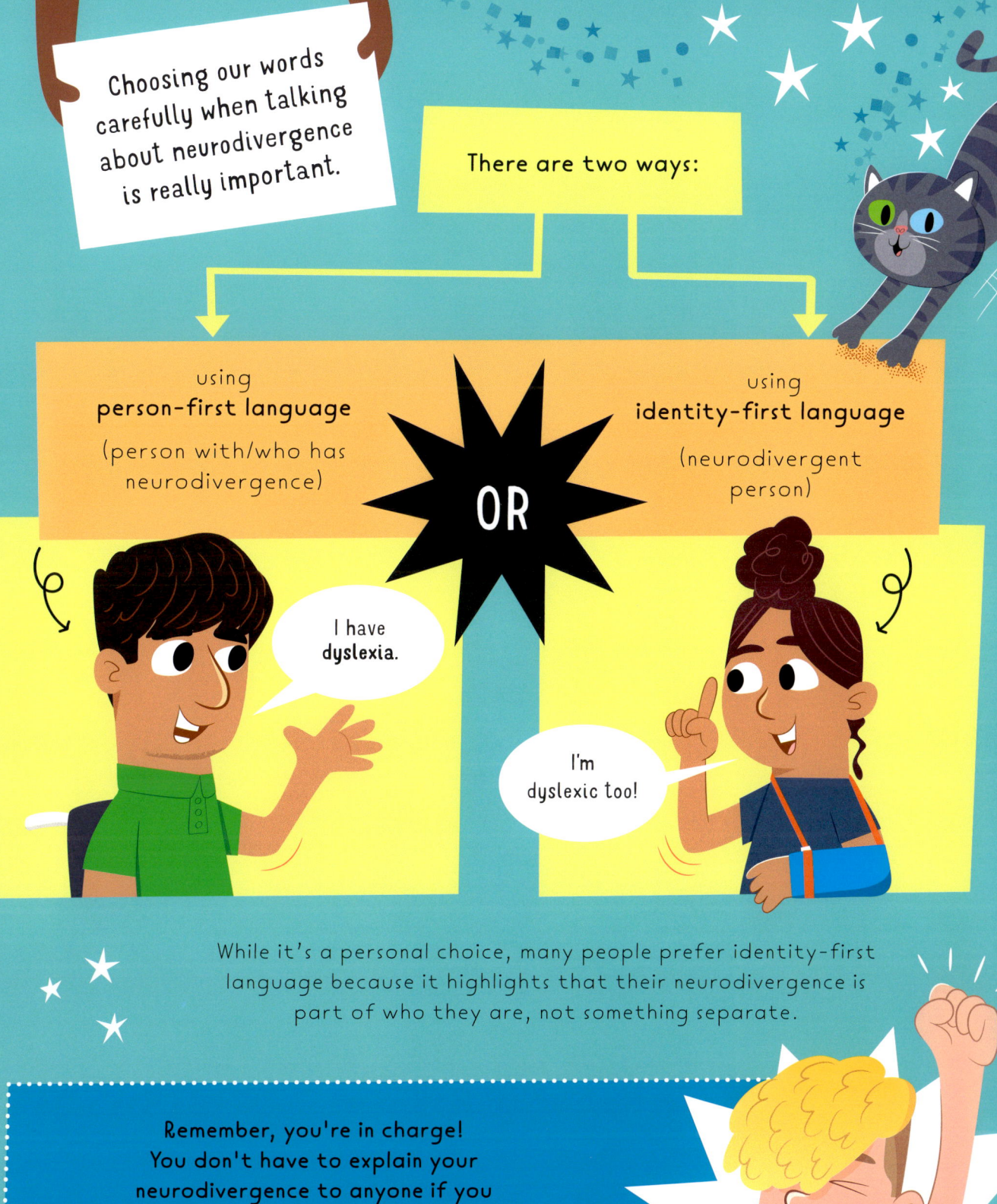

MENTAL HEALTH MATTERS

In different places such as at home, in school, and around your community, you might hear conversations about "mental health." This change is incredibly positive because in the past, people didn't openly talk about this as much as they do today.

Mental health is about how we feel and what we think on the inside. We now know that taking care of our mental health (our thoughts and feelings) is just as important as looking after our physical health (caring for our bodies).

Our mental health can feel like a rollercoaster ride. Sometimes we're really happy, calm, and in control of our emotions. But there are times when things get tough and we might be sad, angry, or confused. Experiencing these ups and downs is a natural and normal part of life.

However, there can be times when we face extreme difficulties, such as stress (feeling overwhelmed), anxiety (feeling very worried), or depression (feeling very sad and hopeless for a long time). This is when it's important to seek support and remember that things can get better.

It's always okay to ask for help if you're struggling with your mental health. Talking to a trusted adult, such as a parent or teacher, can make a big difference. If needed, they can connect you with a mental health professional—someone whose job is to support you.

There are lots of simple, fun things we can do to keep our minds healthy and happy. Spending time with family and friends, doing activities you enjoy, and spending time in nature are all wonderful ways to take care of our mental health.

MINDFUL MAINTENANCE

Your brain is incredible, but it needs your help to take care of it. You've only got one brain, so let's talk about how to keep it in tip-top shape!

Feed your brain
Just like your body, your brain works best on a balanced diet. Foods like fruits and vegetables give it essential vitamins and **nutrients**, while carbohydrates supply it with energy. Proteins help your brain make important **molecules**, such as **neurotransmitters** and **hormones**.

Hydrate your brain
Drinking water all through the day is super-important for a healthy brain. It's advised that children aim to drink about 6-8 cups of water a day.

Protect your brain
Wearing a helmet when biking, skating, or playing sports is like giving your brain its very own shield. It keeps your brain safe, guarding against injuries such as bumps and bruises and reducing the risk of serious harm.

Rest your brain
Sleep is your brain's best friend! It helps your brain recharge and get ready for the next day by sorting and storing all your adventures—like writing it all down in a diary. And, don't forget, it's essential to give your brain a break from screen time too.

Exercise your brain
Moving your body releases happy **hormones**, keeps your heart healthy, and gives your brain a boost. Everyone moves in their own way, so find what makes you feel great!

Listen to your brain
Your brain sends and receives important signals about when you need to take care of your body, like when you have to use the toilet or when you're tired. Pay attention to feelings of pain or discomfort—your brain knows what's best for you!

WHAT CAN YOU DO?

You now know that all brains are wonderful . . .
but not everyone may realize it!
What can you do to help change that?

Celebrate differences . . .

Tell your friends how amazing it is that everyone's brain works in its own way. Encourage them to share what makes their brain awesome!

Be a brain buddy
If someone's struggling with something that you find easy, offer to help them out. It could be with a subject in school or even a new skill. Show them that we all have different strengths.

Listen and learn
When someone talks about their experiences or feelings, be a good listener. You might learn something new about how their brain works and what makes them unique.

Be curious and respectful
It's okay to be curious! Ask questions respectfully to understand different experiences and **perspectives**.

Question how you think...

When you notice your brain using a **stereotype**, question why. Challenge those assumptions and thoughts about how someone is or should be.

Stand up against bullying
No one should be bullied for how their brain works. Speak up, if it's safe to do so, when you see or hear someone being treated unfairly.

Use positive language
Words have power. Use positive and encouraging words to uplift others.

Practice patience
Remember, everyone's brains work at their own pace. Imagine being in their position, and try to be patient and supportive.

Lead by example...

Show people the power of acceptance and understanding through your actions. Others will follow your lead.

Educate and spread understanding
Share your knowledge about the brain and neurodiversity with others to help create a kinder and more **empathetic** community.

GLOSSARY AND USEFUL WORDS

Attention deficit hyperactivity disorder (ADHD)
a neurodevelopmental condition (to do with how the brain develops) that affects some people in different ways, including their attention, self-control, and energy levels

Autism, also known as autism spectrum disorder (ASD)
a neurodevelopmental condition (to do with how the brain develops) that affects some people in different ways, including their communication, social interaction, and perception of the world

Concussion
a type of brain injury caused by an impact, such as a bump or blow to the head or whiplash (a neck injury caused by sudden, forceful movement). It can cause confusion, blurry vision, and memory loss.

Diagnosis (plural—diagnoses)
identifying what medical condition someone has by checking their body for clues (symptoms), doing tests, and asking them about their experiences

Disability (plural—disabilities)
not having equal opportunities in everyday life because of a physical or social barrier

Discrimination
treating people unfairly because of their particular characteristics, such as their disability, gender, or background

Dyscalculia
a specific learning difficulty where someone struggles with understanding and working with numbers

Dysgraphia
a specific learning difficulty where someone finds it challenging to write clearly or express their thoughts effectively through written words

Dyslexia
a specific learning difficulty that affects a person's reading and spelling of words, as well as their ability to understand language and process information

Dyspraxia, also known as developmental coordination disorder (DCD)
a specific learning difficulty that affects someone's physical coordination and muscle movement

Empathetic
being able to recognize, understand, and share the feelings of others

Hormones
chemicals produced by our bodies that control things like how we grow, our emotions, and the changes that happen during puberty

Hyperlexia
a condition where a person demonstrates an advanced and exceptional ability to read well at a very young age, but may still have trouble understanding the meaning of the written words

Molecules
tiny particles that make up everything in the world, but are too small to be seen with the naked eye

Neurodivergent
someone whose ways of thinking are different to what society sees as "typical," including people with neurodevelopmental conditions or specific learning difficulties

Neuroinclusivity
helping neurotypical people to learn about neurodivergence and creating environments where everyone feels welcomed, comfortable, and supported to be themselves

Neurons
cells in the brain that send and receive messages that allow you to think, move, and feel

Neurotransmitters
chemical messengers in the brain that help neurons communicate with each other and with other important cells such as those in our muscles

Neurotypical
someone whose ways of thinking are seen as expected by society

Nutrients
vitamins, proteins, and other essential substances that are found in food and drinks and help our body to grow and be healthy

Obsessive compulsive disorder (OCD)
a mental health condition where a person has thoughts and behaviors that they feel they must do to ease their anxiety or discomfort

Perspectives
different ways of looking at or understanding something

Resilience
the ability to bounce back from challenges or difficulties, showing strength and determination, and continuing to move forwards

Sexual orientation
who someone is or isn't attracted to romantically or sexually

Stereotype
a fixed idea or belief about a person or group that may not be accurate or true

Synesthesia
a condition that causes two or more senses that are usually triggered separately to become linked. For example, hearing music may trigger a specific sight, taste, or smell.

Tics
fast and repetitive movements or sounds, such as blinking, twitching, or making noises, that some people make due to strong urges or because they can't control the action

Tourette's syndrome
a neurodevelopmental condition (to do with how the brain develops), named after a neurologist. It causes some people to make sounds and movements called tics.